JUPITER

Robin Kerrod

Lerner Publications Company • Minneapolis

This edition published in 2000

Lerner Publications Company.
A Division of Lerner Publishing Group
241 First Avenue North, Minneapolis MN 55401

Website address: www.lernerbooks.com

© 2000 by Graham Beehag Books

Library of Congress Cataloging-in-Publication Data

Kerrod, Robin
  Jupiter / Robin Kerrod.
    p. cm.
  Includes index.
  Summary: An introduction to the planet Jupiter, with
information about its powerful gravity, stormy atmosphere,
and various moons.
  ISBN 0-8225-3907-1 (lib. bdg.)
  1. Jupiter (Planet)—Juvenile literature. [1. Jupiter (Planet)]
I. Title. II. Series: Kerrod, Robin. Planet library.
QB661.K47 2000   99.040069
523.45—dc21

Printed in Singapore by Tat Wei Printing Packaging Pte Ltd
Bound in the United States of America
1 2 3 4 5 6 – OS – 05 04 03 02 01 00

CONTENTS

JUPITER

4

Introducing Jupiter

Jupiter is the largest of the nine planets in the solar system—the family of bodies that circle in space around the Sun. Jupiter is so big that all of the other planets could fit inside it easily. It has more than 300 times the mass, or matter, of our own planet, Earth.

Jupiter is quite a different kind of planet from Earth. Earth is one of the four rocky planets of the inner solar system—Mercury, Venus, Earth, and Mars. Jupiter is one of the gas giants of the outer solar system—Jupiter, Saturn, Uranus, and Neptune. The ninth planet, Pluto, is a small ball of rock and ice.

Jupiter is one of the brightest objects in our night sky, after the Moon and the planet Venus. Like all the planets, Jupiter shines because it reflects light from the Sun. In powerful telescopes, we see that Jupiter is a colorful planet, with light and dark bands and white and colored spots. These are features of Jupiter's atmosphere—the thick layer of gases that covers the planet.

Jupiter is the center of its own kind of "solar system." At least 16 moons circle around Jupiter, in much the same way the planets circle around the Sun. Jupiter keeps its large family of moons in place with its enormous gravity. Gravity is the attraction, or pull, that a heavenly body has on objects on or near it. Jupiter's moons have an amazing variety of features.

Left: Jupiter is a colorful planet, marked by bands of fast-moving clouds. Its most prominent feature is a huge oval region called the Great Red Spot, seen here at bottom left. On the right side is Io, the nearest of Jupiter's four large moons.

Right: An artist's impression of the Great Red Spot, seen from Amalthea, a tiny moon that orbits much closer to Jupiter than Io.

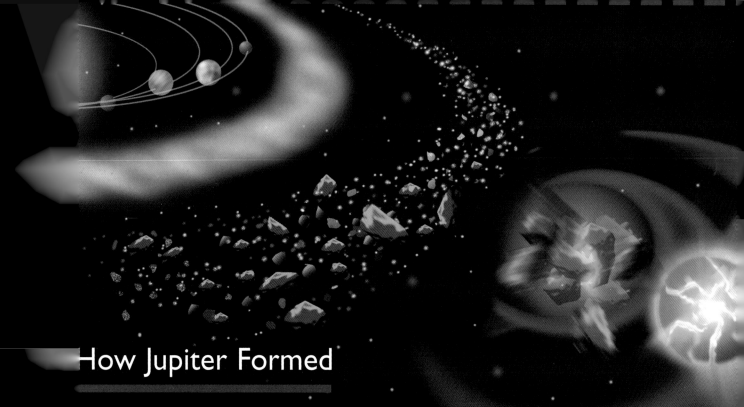

How Jupiter Formed

Jupiter was born about 4.6 billion years ago, at the same time Earth and the other planets formed. It became a different kind of planet from Earth because it formed so much farther away from the Sun.

The solar system formed out of a huge cloud of dust and gas, mainly hydrogen and helium. Over time, the cloud shrank into a huge spinning ball, with a disk of gas and dust circling around it.

The ball of gas at the center gradually became smaller as it collapsed under the pull of gravity between its particles. As the ball collapsed, it heated up. It eventually started to glow, and in time it began to shine as a star, the Sun.

As the new Sun shone, it blasted surrounding layers of gas away from it in the form of a furious solar wind. This process continued for millions of years.

PLANETS FORM

Meanwhile, changes were taking place in the disk spinning around the Sun. The inner part of the disk was quite warm. There, lumps of rock and metal were forming out of smaller bits as they kept bumping into one another. In time, these lumps became bigger and bigger until they formed into the rocky Earth-like planets of the inner solar system.

Farther out, the disk was much colder. The bits of matter found there were lumps of ice and frozen gases. Over time, these lumps also grew into large bodies—the

All the while, the Sun had been blowing gases away from the inner part of the disk into the colder outer part. These gases, mainly hydrogen and helium, formed a great cloud around the icy bodies that had formed there. These bodies began to attract more and more gases, growing bigger in the process. The body we call Jupiter happened to be in the thickest part of the gas cloud, and it attracted the most gas and eventually became the largest planet.

Stages in the formation of Jupiter. From left to right: The young Sun blows gas from the inner part of the solar system. Beyond the asteroid belt, lumps of matter form into another planet. Gases gather around it, and it grows into Jupiter as we know it.

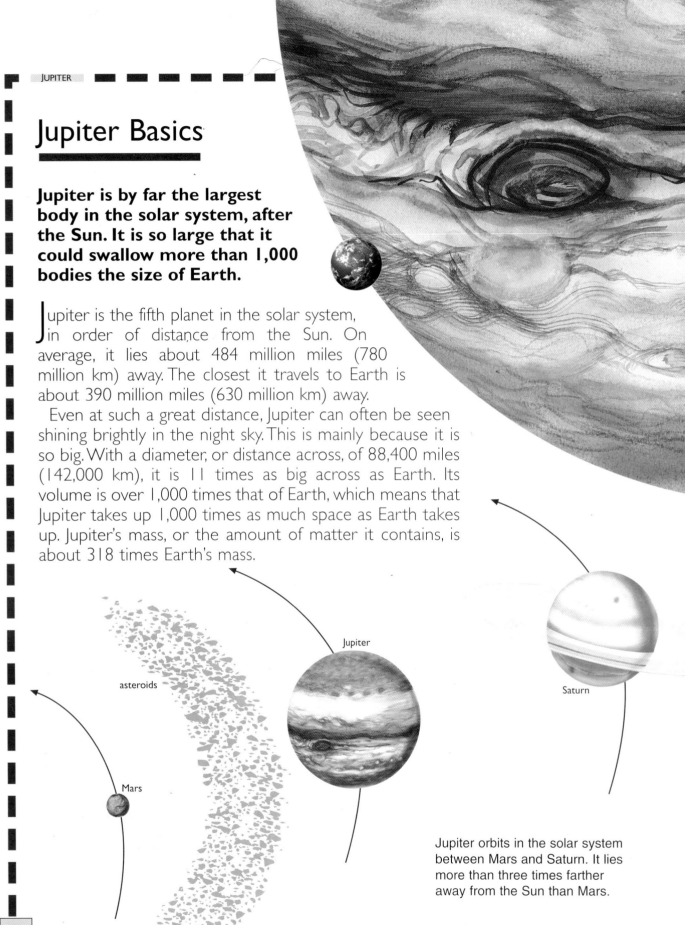

Jupiter Basics

Jupiter is by far the largest body in the solar system, after the Sun. It is so large that it could swallow more than 1,000 bodies the size of Earth.

Jupiter is the fifth planet in the solar system, in order of distance from the Sun. On average, it lies about 484 million miles (780 million km) away. The closest it travels to Earth is about 390 million miles (630 million km) away.

Even at such a great distance, Jupiter can often be seen shining brightly in the night sky. This is mainly because it is so big. With a diameter, or distance across, of 88,400 miles (142,000 km), it is 11 times as big across as Earth. Its volume is over 1,000 times that of Earth, which means that Jupiter takes up 1,000 times as much space as Earth takes up. Jupiter's mass, or the amount of matter it contains, is about 318 times Earth's mass.

asteroids

Jupiter

Saturn

Mars

Jupiter orbits in the solar system between Mars and Saturn. It lies more than three times farther away from the Sun than Mars.

Left: Jupiter is truly a giant of a planet, dwarfing our own planet Earth. Several Earths could fit into Jupiter's huge permanent storm region we know as the Great Red Spot.

Right: Photographs taken through telescopes on Earth show the multicolored bands of Jupiter. Various spots also show up, which are the centers of severe storms.

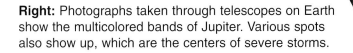

STAR POINT

Though Jupiter is the biggest planet, it rotates in space 15 times faster than the smallest planet, Pluto.

SPINNING AROUND

As Jupiter travels around the Sun, it also moves in another way. Like all the planets, Jupiter rotates, or spins around on its axis. An axis is an imaginary line through an object from its north pole to its south pole.

Jupiter rotates faster than any other planet. In fact, this huge planet spins all the way around once in less than 10 hours. This is less than half the time Earth takes to spin around once, which is 24 hours, or 1 day. Because Jupiter spins so fast, it bulges out at the middle around its equator—the imaginary line around the center of a planet, midway between its north and south poles. This fast spinning also causes Jupiter to be slightly flattened at its poles.

JUPITER DATA

Diameter at equator:
88,400 miles (142,000 km)

Average distance from Sun:
484,000,000 miles (780,000,000 km)

Rotates in: 9 hours, 50 minutes

Orbits Sun in: 11.9 years

Moons: 16 known

JUPITER'S MAKEUP

Jupiter's makeup is very different from that of a rocky planet like Earth. Earth is made up of three main layers of rock, with a metal core at the center. Jupiter is made up of layers too, but mostly of gas and liquid gas.

When we look at Jupiter through a telescope, we can see various colored markings, including light bands and dark bands. These markings are actually the clouds of Jupiter's atmosphere, which is its top layer. The temperature here is very cold—about –200°F (–130°C).

Below the atmosphere is Jupiter's surface. The planet's surface is not solid, but instead is a vast ocean of liquid hydrogen. In this part of Jupiter, the atmospheric pressure—the force of the gases pressing down—is enormous. Deeper down inside Jupiter, the pressure is so great that it forces the hydrogen to turn into a metallic liquid. Temperatures here may rise as high as 43,000°F (24,000°C). At the center of the planet is probably a core of rock and ice, with about 10 to 20 times the mass of Earth.

atmosphere

liquid hydrogen

liquid metallic hydrogen

rocky core

This look inside Jupiter shows the various layers that make it up. The outer atmosphere of gases contains mainly hydrogen and helium. Deeper down, immense pressure changes the hydrogen gas first into liquid and then into a kind of liquid metal.

Stormy Atmosphere

Clouds and furious winds rush through Jupiter's atmosphere. Lightning flashes among the clouds, as storms spring up all over the planet.

Hydrogen and helium are the two main gases in Jupiter's stormy atmosphere. Traces of other gases are found there as well. They include ammonia, water vapor (water in the form of a gas), and methane. Methane is the main gas in the natural gas used to heat homes on Earth. Gases containing sulfur are also found in Jupiter's atmosphere.

The colored bands in Jupiter's atmosphere are actually layers of clouds. The clouds form into bands because the atmosphere moves so quickly. Astronomers call the pale bands zones and the darker bands belts. The main colors in the atmosphere are white and red. The white regions are high clouds of tiny ammonia crystals. The reddish ones are lower clouds made up of sulfur compounds.

Jupiter's main visible features are the dark and light bands, which astronomers call belts and zones. They have each been given names so that astronomers can refer to them when reporting the changes taking place on the planet.

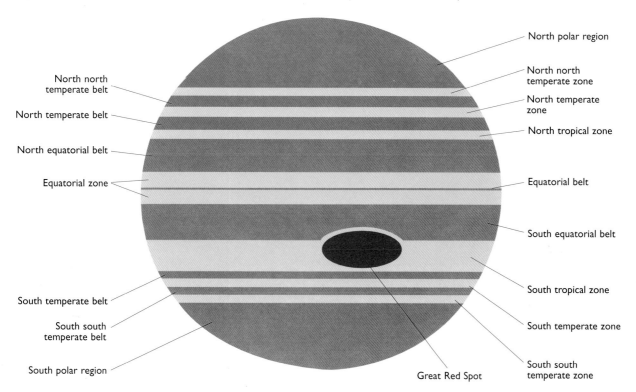

North polar region

North north temperate zone

North temperate zone

North tropical zone

Equatorial belt

South equatorial belt

South tropical zone

South temperate zone

South south temperate zone

Great Red Spot

South polar region

South south temperate belt

South temperate belt

Equatorial zone

North equatorial belt

North temperate belt

North north temperate belt

In the belts and zones, the winds blow strongly, at speeds of up to 300 miles (500 km) an hour. They do not all blow in the same direction. Some blow toward the east, in the same direction the planet rotates. Other winds blow toward the west.

STORMY WEATHER

Where the winds on Jupiter blowing in opposite directions meet, the atmosphere swirls around furiously. Great wavy patterns, or eddies, form, as do white and colored spots. In these spots, the atmosphere whirls around like in a gigantic whirlwind. Lightning flashes between the clouds as it does in thunderstorms on Earth.

Above: The appearance of the belts and zones on Jupiter changes hour by hour as furious winds rage in the atmosphere.

LIFE AMONG THE CLOUDS?

The deeper one goes into Jupiter, the higher the temperature. Beneath the cold upper atmosphere are warm layers of gas. In the warm layers, some scientists think that chemical reactions might take place between the gases of the atmosphere, such as ammonia, methane, and water. Scientists believe that long ago these gases might have combined to form simple living things. This is probably what happened on Earth billions of years ago.

Some of the stormiest regions on Jupiter occur around the Great Red Spot. Here the clouds form fascinating wavy patterns as the winds swirl and eddy.

On Jupiter, say scientists, it is possible that special life forms spend their whole lives floating in the warm layers of the atmosphere.

THE GREAT RED SPOT

The Great Red Spot is Jupiter's biggest feature. Astronomers had noticed this reddish oval region in Jupiter's atmosphere and wondered what it was for more than 300 years. Space probes have shown that the Great Red Spot is an enormous storm, much like a hurricane on Earth.

The Great Red Spot measures about 25,000 miles (40,000 km) across. Three bodies the size of Earth could fit in it side by side. The spot changes its size and color slightly over time. But its position on Jupiter never changes.

STAR POINT

The Great Red Spot spins around once about every six days.

This close-up of the Great Red Spot was taken by a *Voyager* space probe. The gases in this great storm region circulate in a counterclockwise direction. The color of the Spot seems to come from the presence of the chemical element phosphorus.

Jupiter's Magnetism

Jupiter is a magnetic planet, like Earth. But its magnetism is much more powerful and reaches out millions of miles into space.

All around Earth is an invisible force called magnetism. Earth's magnetism is produced in its core, which is made up mainly of hot liquid iron. When Earth rotates, the movement creates currents of electricity in the core.

Jupiter has magnetism too, and astronomers believe Jupiter's magnetism is produced in much the same way Earth's is—by electric currents in liquid metal. But on Jupiter, the liquid metal

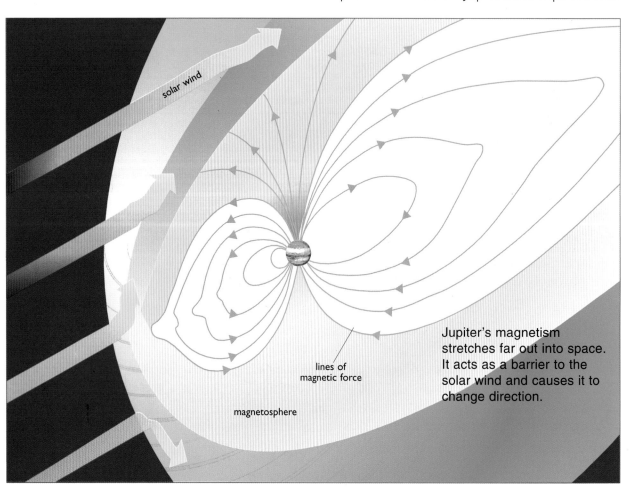

solar wind

lines of
magnetic force

magnetosphere

Jupiter's magnetism stretches far out into space. It acts as a barrier to the solar wind and causes it to change direction.

is not found in an iron core. The source of Jupiter's magnetism is the metallic hydrogen found in the layer surrounding Jupiter's core.

THE MAGNETOSPHERE

Jupiter's magnetism is 20,000 times as strong as Earth's. It acts all around the planet and reaches out millions of miles into space. The magnetic region is called the magnetosphere. It is constantly moving in space because it rotates with Jupiter.

The magnetosphere does not stretch evenly in all directions around Jupiter. This is because of the solar wind—the stream of particles given off by the Sun. On the side of Jupiter facing the Sun, the magnetosphere is squashed somewhat by the solar wind.

Some of the particles in the solar wind flow around the magnetosphere. Some get trapped inside it. Others travel in toward Jupiter and enter its atmosphere. They make the gases in the atmosphere glow. Something similar happens on Earth when particles from the solar wind enter Earth's atmosphere and make it glow. We call these shimmering lights the aurora, or the Northern and Southern Lights.

RADIO JUPITER

Unlike other planets, Jupiter gives off strong radio radiation. Astronomers pick up these waves with radio telescopes on Earth. Most astronomers believe the radio radiation is caused by particles from the solar wind whizzing around in Jupiter's magnetosphere. Other radio waves are produced by powerful flashes of lightning during the great storms that often take place in Jupiter's atmosphere.

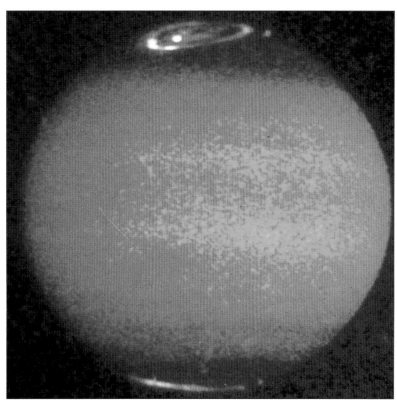

Jupiter's "northern lights" show up in this picture of the planet taken by the Hubble Space Telescope. They are displays of aurora around the planet's north polar regions.

Powerful Gravity

Jupiter has an extremely powerful gravitational pull. It keeps a large family of moons in place, and it affects other nearby bodies in space.

If you could live on Jupiter, you would find it difficult to walk. This is because Jupiter's gravity is more than two and a half times more powerful than Earth's gravity. That means that if you weigh 100 pounds on Earth, you would weigh 264 pounds on Jupiter.

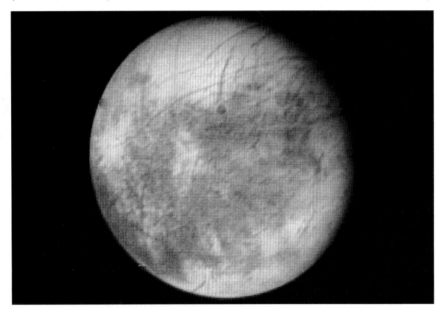

Europa is one of the 16 moons that Jupiter holds on to with its powerful gravity. In the solar system, only the Sun has a stronger gravitational pull than Jupiter.

Galilean moons

Jupiter

inner family

outer family

middle family

Jupiter is at the center of its own miniature "solar system," formed by its 16 circling moons. They fall into four families, according to their orbits. From one side to the other, Jupiter's "solar system" spans a distance of nearly 30 million miles (50 million km).

Every body in the universe has gravity due to its mass. The more a body's mass, the more powerful its gravitational pull is. Jupiter has a powerful pull because it is so big.

Jupiter's gravity reaches out a long way into space. It is the force that keeps Jupiter's 16 moons circling around the planet. Even 15 million miles (25 million km) away, Jupiter's gravity is strong enough to hold on to a tiny rocky moon called Sinope, which is only about 22 miles (35 km) across.

JUPITER AND THE ASTEROIDS

Jupiter's gravity also affects some of the asteroids. Asteroids are rocky bodies that orbit the Sun like small planets. Most asteroids are found between the orbits of Mars and Jupiter. But Jupiter has captured two groups of asteroids called the Trojans. They travel around the Sun in the same orbit as Jupiter. One group travels in front of Jupiter, and the other travels behind it.

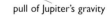
pull of Jupiter's gravity

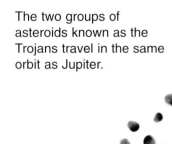
The two groups of asteroids known as the Trojans travel in the same orbit as Jupiter.

JUPITER AND COMETS

Jupiter's gravity may also affect comets when they travel near the planet. Comets are tiny balls of ice and dust that orbit the Sun. They travel in toward the Sun from the outer parts of the solar system. Afterward, they usually travel back to where they came from. Most comets take thousands of years to make the journey in toward the Sun then back out again.

If a comet passes near Jupiter, the giant planet's gravity pulls it and changes its path. This may force a comet into a much shorter orbit so that it travels back to the Sun in a matter of years rather than thousands of years. Several comets travel between the Sun and Jupiter, and astronomers say that they belong to Jupiter's comet family.

COMET COLLISIONS

Jupiter's powerful gravity sometimes causes a passing comet to break apart. This happened to a comet called Shoemaker-Levy 9 in 1992. In March 1993, astronomers discovered that pieces of the comet were traveling toward Jupiter. In July 1994, the pieces smashed into Jupiter. Some of the collisions created huge fireballs. The collisions left markings in Jupiter's atmosphere that lasted for days. The largest was almost as big as Earth.

Top: Many comets are affected by Jupiter's powerful gravity. When this happens, they may become regular visitors to our skies, like Halley's comet, shown here.

Left: In the summer of 1994, a string of fragments of the comet Shoemaker-Levy 9 made a beeline for Jupiter. When they crashed into the planet, they left prominent "scars" in the atmosphere (inset picture).

JUPITER'S RINGS

Until 1979, astronomers knew of only two planets—Saturn and Uranus—that had rings around them. But in 1979 the space probe *Voyager 1* spotted a system of very thin, faint rings when it flew past Jupiter. This was a complete surprise. In 1984, astronomers discovered that Neptune has rings, too.

The main part of Jupiter's ring is located about 76,500 miles (123,000 km) from the planet, and it circles the planet's equator. The ring is only about 4,000 miles (6,000 km) across and about 19 miles (30 km) thick. Smaller rings circle inside and outside this main ring.

The rings are not solid, like metal washers. They are made up of tiny particles of rock. These particles travel in orbit around Jupiter at high speed. They appear to us as rings because they move so fast that the light they reflect is blurred.

STAR POINT

Particles in Jupiter's rings whiz around the planet in only about six hours.

Part of the ring system around Jupiter, photographed by the *Voyager 1* space probe, which discovered it (right). It shows up more clearly in a picture of the planet taken with ultraviolet light (below).

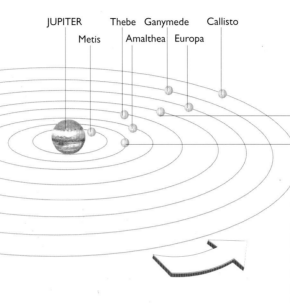

JUPITER   Thebe Ganymede   Callisto

Metis   Amalthea   Europa

Io

Adrastea

The scattered family of Jupiter's moons. The four moons of the distant outer family circle the planet in the opposite direction from the other moons.

Jupiter's Moons

Jupiter's moons are some of the most fascinating bodies in our solar sytem. Most are tiny, but four are planet-sized.

The Italian astronomer Galileo was first to train a telescope on the heavens and first to spot Jupiter's four large moons.

From Earth, we can observe 13 of Jupiter's 16 known moons. The other three were discovered by the *Voyager* space probes in 1979 and 1980. Jupiter's four largest moons circle quite close to the planet. They are Io, Europa, Ganymede, and Callisto. These are called the Galilean moons, after the Italian astronomer Galileo. He discovered them with a small homemade telescope in 1610.

The Galilean moons can easily be seen from Earth. Through binoculars, you can see them lined up around Jupiter. But it wasn't until the *Voyager* probes visited Jupiter in 1979 that we received detailed pictures and information about Jupiter's moons. Astronomers were surprised to learn how much the moons differ from one another.

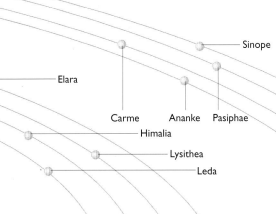

Sinope

Elara

Carme   Ananke Pasiphae

Himalia

Lysithea

Leda

Right: This montage of *Voyager* pictures shows Jupiter and the four Galilean moons. From top to bottom, they are Io, Europa, Ganymede, and, in the right-hand corner, Callisto.

THE DWARF MOONS

While the Galilean moons measure thousands of miles across, most of Jupiter's other moons are much smaller. The smaller moons range in size from 9 to 106 miles (15 to 170 km) in diameter. Metis, the moon closest to the planet, measures only about 25 miles (40 km) across.

Four of Jupiter's tiny moons circle the planet 10 times farther out than Callisto, the farthest out of the Galilean moons. They circle Jupiter at a distance of more than 7 million miles (11 million km). Another group of four moons circle the planet at about twice this distance. Jupiter's most distant moon, Sinope, takes more than three Earth-years to circle around Jupiter.

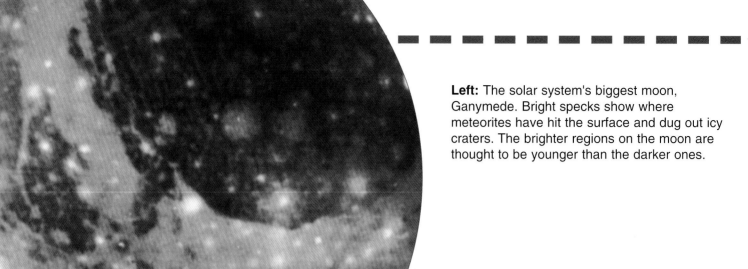

Left: The solar system's biggest moon, Ganymede. Bright specks show where meteorites have hit the surface and dug out icy craters. The brighter regions on the moon are thought to be younger than the darker ones.

Below: Callisto is the most heavily cratered of Jupiter's moons. It probably has the oldest surface.

GANYMEDE

Ganymede is Jupiter's largest moon. It is also the largest moon in the entire solar system. With a diameter of 3,278 miles (5,276 km), it is two and a half times as large as our own Moon, and it is larger than the planet Mercury.

Scientists believe that Ganymede has three layers. In its center is a small core of iron or iron and sulfur. Surrounding the core is a rocky mantle, and on top is an icy surface layer, or crust. Ganymede's surface has dark rocky areas and white icy areas. Craters created by meteorites cover much of the dark areas. Meteorites are lumps of rock that fall from space onto a planet or moon's surface.

Ganymede's icy areas are covered by dark grooves, or valleys, which lie between long, narrow ridges. Astronomers believe these grooves and ridges were caused by movements in Ganymede's crust long ago.

CALLISTO

Callisto is Jupiter's second-largest moon. Measuring 2,995 miles (4,820 km) across, it is nearly the same size as the planet Mercury. Like Ganymede, it is also made up of rock and ice.

But unlike Ganymede, Callisto is completely covered with craters. It has more craters than any other moon in the solar system. Most of these craters were formed billions of years ago. Callisto has no mountains and valleys like those found on Ganymede. So astronomers believe that there have been no movements in Callisto's crust for billions of years.

EUROPA

Europa looks very different from the other Galilean moons. Completely covered with ice, Europa's surface is light in color and very smooth. The icy surface probably formed when liquid water seeped up from below and froze.

A network of dark grooves and ridges can be seen all over Europa. The dark grooves are probably cracks in the icy crust. They extend up to thousands of miles across the surface. Scientists believe that an ocean of liquid water may still lie under the icy surface.

STAR POINT

Ice-covered Europa is the smoothest body we know in the solar system.

Above: Few craters are visible on the smooth surface of Europa. It is the smallest of the Galilean moons, with a diameter of 1,942 miles (3,126 km).

Right: Europa has a network of dark lines, which are probably cracks in the icy surface that have filled with dirty material from below.

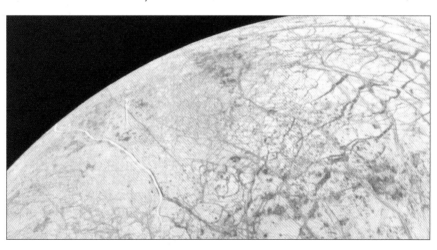

Io, the Pizza Moon

Io is the Galilean moon that is closest to Jupiter, and it is a very unusual moon. Io is the only body in the solar system, other than Earth, with active volcanoes. Its brightly colored markings are caused by volcanoes erupting on its surface.

With a diameter of 2,255 miles (3,630 km), Io is slightly larger than our own Moon. While most other moons are dull in color, Io is mainly yellow and orange, with pale and dark markings. It has been called the pizza moon because its surface looks somewhat like the colorful surface of a pizza.

The astonishing surface of Io (top) owes its color to the sulfur that spews out of its many volcanoes. Vapor and dust erupting from the volcanoes shoot high above the surface (below).

Sulfur Landscape

The volcanoes on Io are not like the volcanoes on Earth. On Earth, volcanoes are places where red-hot liquid rock under the ground forces its way up to the surface. The rock then quickly cools and becomes hard.

On Io, the substance that forces its way to the surface in volcanoes is sulfur. Sulfur is yellowish-orange in color, which explains the color of Io's surface. Sulfur is found around volcanoes on Earth in small amounts. On Io, sulfur from volcanoes has covered the whole surface.

Io's volcanoes also shoot large amounts of sulfur vapor into the sky. This freezes to form particles of a white substance that falls like snow on the surrounding landscape.

How the Volcanoes Form

The volcanoes on Io are almost certainly caused by Jupiter's enormous gravity. It pulls different parts of Io at different times as the moon spins around in space. This sets up movements inside Io, which make it heat up. The heat melts the sulfur and forces it out through cracks in the hard outer crust.

In this *Voyager* picture, Io is seen over Jupiter's famous Great Red Spot. The other moon in the picture is Europa.

Linda's Luck

The first volcano on Io was discovered by Linda Morabito of the Jet Propulsion Laboratory in California. She found it in February 1979 when studying photographs of Jupiter and its moons taken by the *Voyager 1* space probe.

Missions to Jupiter

In the 1970s space probes from NASA (National Aeronautics and Space Administration) gave us our first close look at Jupiter and its moons. They sent back amazing pictures and vast quantities of new information.

Pioneer 10 was a true pioneer of deep space exploration, becoming the first space probe to navigate the asteroid belt and experience the intense radiation in Jupiter's magnetosphere. *Pioneer 11* followed a different path around Jupiter so that it could carry on to Saturn.

In March 1972 a powerful Atlas-Centaur rocket blasted off the launch pad at Cape Canaveral in Florida. It was carrying a space probe called *Pioneer 10*. Faster and faster the rocket sped into the air and then into space. When its engines stopped firing, the space probe separated and set off on a journey that would take it to Jupiter. The probe left Earth traveling at a speed of more than 32,000 miles (51,000 km)

an hour. This was the fastest any vehicle had ever traveled—more than 15 times faster than a rifle bullet.

The biggest part of *Pioneer 10* was the dish antenna, which measured about 9 feet (2.7 m) across. Behind it was the main body of the spacecraft, which included a box containing electronic equipment and various information-gathering instruments.

DODGING THE ASTEROIDS

To reach Jupiter, *Pioneer 10* had to pass through the asteroid belt. In this region, thousands of asteroids circle the Sun like miniature planets. No one knew for sure whether the spacecraft could get through the belt without being hit. But it did, and it flew past Jupiter in December 1973.

As it passed by, *Pioneer* took photographs of Jupiter. They showed that the Great Red Spot is a huge weather system. *Pioneer*'s instruments showed that Jupiter has a powerful magnetic field and gives off radiation. Scientists also learned that the planet gives off more heat than it receives from the Sun.

REPEAT PERFORMANCE

After traveling past Jupiter, *Pioneer 10* sped off into interplanetary space. But it continued to send back information. *Pioneer 11* was launched in April 1973 and it reached Jupiter in December 1974. It zoomed in to photograph the planet's south pole. Then, after swinging around Jupiter, it began to follow a long, looping path that would take it to Saturn.

STAR POINT

NASA scientists broke off contact with *Pioneer 10* in April 1997, when the spacecraft was more than 6 billion miles (10 billion km) from Earth.

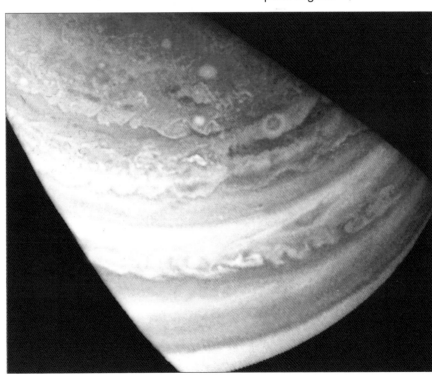

Pioneer 11 showed in detail the vigorous activity taking place in Jupiter's atmosphere for the first time. This view shows the planet's north polar region.

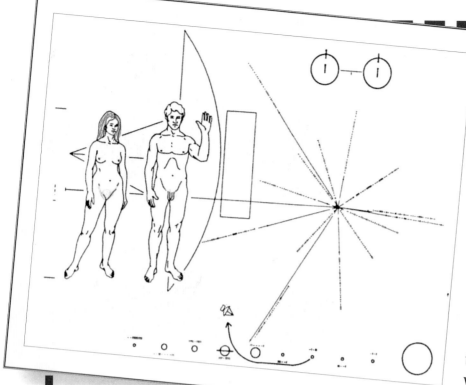

Pictures for Aliens

After leaving the solar system, the *Pioneer* probes will venture into interstellar space—the space between the stars. One day, perhaps, intelligent aliens somewhere in space may come across them and wonder where they came from. To give them an answer, each *Pioneer* spacecraft carries a metal plaque (shown here) with a message in pictures. Created by astronomer Carl Sagan, it tells where the spacecraft came from and who sent it.

DEEP SPACE VOYAGERS

The *Pioneer* probes proved that spacecraft could safely travel through the asteroid belt and send pictures and information over vast distances. So NASA decided to send two more probes to Jupiter and beyond. Named *Voyager 1* and *Voyager 2*, they were launched in the summer of 1977.

The *Voyagers* had a dish antenna about 12 feet (3.7 m) across. Most of its instruments were carried on a movable platform so that they could easily be pointed in different directions. Its cameras were much better than those on the *Pioneers*.

ASTOUNDING IMAGES

Voyager 1 made its closest approach to Jupiter in March 1979. But long before that it was sending back pictures that were astounding mission scientists. They showed the swirling atmosphere of the giant planet in great detail. Scientists saw clouds racing along at high speeds and storms breaking out all over Jupiter.

Voyager also sent back pictures of Jupiter's moons. Every

Voyager 2 blasts off the launch pad at Cape Canaveral on August 20, 1977, about two weeks before its sister craft, *Voyager 1*. But *Voyager 1* reached Jupiter first.

moon they spied looked different, which came as a big surprise. Oddest of all was the brilliantly colored Io, nicknamed the pizza moon, which had volcanoes. Then came another surprise, when *Voyager 1* spotted Jupiter's ring.

Soon *Voyager 1* was moving on to its next target, Saturn. *Voyager 2* took its place, flying closest to the planet in July 1979. It sent back more information and more amazing pictures of storms, moons, volcanoes, and rings. In just a few months, the two *Voyager* spacecraft had revolutionized our knowledge of the biggest planet in our solar system.

STAR POINT

Voyager 2 went on to visit three more planets after Jupiter (Saturn, Uranus, and Neptune) and become the most successful space probe in the history of space exploration.

Left: The *Voyagers* were first to spot the incredible multicolored landscape of Io and its erupting volcanoes. Nothing like it had ever been seen before.

Below: Colorization has been used in this *Voyager* image to pick out the clouds and currents swirling in and around Jupiter's Great Red Spot.

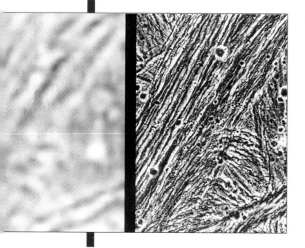

Galileo has much better cameras than the *Voyager* probes, as these images of the same region of Ganymede show. *Galileo* can spot details as small as about 200 feet (70 m).

GALILEO TO JUPITER

NASA's next probe to Jupiter was launched in February 1990 by astronauts on the space shuttle *Atlantis*. When *Galileo* reached Jupiter in December 1995, it spent more than two years exploring the Galilean moons.

Galileo took almost six years to reach Jupiter because it followed a long route through the solar system. It looped twice around the Sun, passing Venus once and Earth twice. Each time it passed one of these bodies, it was speeded up by the body's gravity. It also passed twice through the asteroid belt, taking the first-ever pictures of asteroids.

Just before *Galileo* went into orbit around Jupiter in December 1995, it released a small probe. The probe parachuted into the planet's thick atmosphere and sent back information about the conditions there. Unfortunately, *Galileo*'s main antenna got stuck during the journey to Jupiter, which meant that it could not send back as many pictures as planned. But the pictures that did come back were excellent and showed more details of Jupiter's atmosphere and moons.

This illustration shows the small probe released by *Galileo* parachuting into Jupiter's stormy atmosphere. It radioed information to the main probe, which then relayed it back to Earth.

Glossary

asteroid: a rocky body that orbits the Sun between Mars and Jupiter

asteroid belt: a ring-shaped region in the solar system, between Mars and Jupiter, in which most asteroids are found

atmosphere: the layer of gases around a planet or moon

atmospheric pressure: the force of the gases in an atmosphere pressing down

aurora: a glow produced in the polar regions of some planets by solar wind particles entering the atmosphere

axis: an imaginary line through a planet from its north to its south pole

belt: a dark-colored band in Jupiter's atmosphere

comet: a small body, made up of dust and ice, that orbits the Sun and shines when it nears the Sun

core: the center part of a planet or moon

crater: a pit on the surface of a planet or moon

crust: the hard surface of a rocky or icy planet or moon

equator: an imaginary line around the center of a heavenly body

Galilean moons: the four large moons of Jupiter, discovered by Galileo

gravity: the attraction, or pull, that a heavenly body has on objects on or near it

Great Red Spot: a huge rotating storm system on Jupiter

interplanetary space: the space between the planets

magnetosphere: the region in space around a planet where its magnetism can be detected

mass: the amount of matter in a body

meteorite: a lump of rock or metal from space that hits a planet or a moon

moon: a natural satellite of a planet

NASA: the National Aeronautics and Space Administration, which organizes space activities in the United States

orbit: the path in space of one heavenly body around another, such as Jupiter around the Sun

probe: an unmanned spacecraft that travels from Earth to one or more heavenly bodies

ring system: a set of rings found around a giant planet, made up of fine particles or lumps of rock and ice

solar wind: a stream of particles given off by the Sun

Trojans: groups of asteroids that travel in the same orbit as Jupiter

zone: a light-colored band in Jupiter's atmosphere

Index

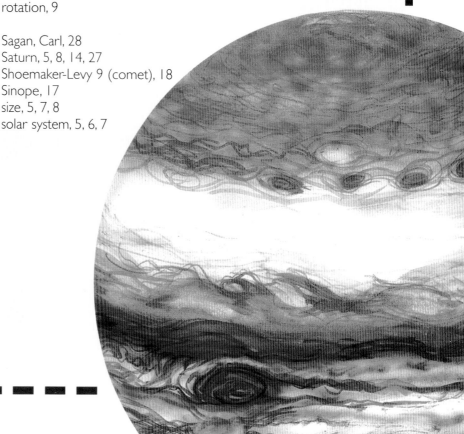